The Pioneer (Donner) Monument

THE ORIGIN OF A STATUE

The Pioneer
(Donner)
Monument
The Origin of a Statue

Doris Foley

NEVADA COUNTY HISTORICAL SOCIETY

NEVADA CITY, CALIFORNIA

PUBLISHED BY THE NEVADA COUNTY HISTORICAL SOCIETY
P.O. BOX 1300, NEVADA CITY, CA 95959

DESIGN, COMPOSITION AND PRODUCTION BY DAVID COMSTOCK
PRINTED AND BOUND BY THOMSON-SHORE, INC.

ISBN 0-915641-10-0

Contents

ILLUSTRATIONS

Acknowledgments

Mary Ashe, Art Department, San Francisco Public Library
Bancroft Library, University of California, Berkeley
Peter T. Conmy, Director of Historical Research, N.S.G.W.
William N. Davis, Jr., Director of California State Archives
Emigrant Trail Museum, Donner Memorial State Park
Gladys Hansen, Special Collections Department, San Francisco
      Public Library
Ellen Jones, Manuscript Curator, Bancroft Library
Debra Luckinbill, granddaughter of Chester W. Chapman
Nevada City Public Library
Searls Historical Library, Nevada County Historical Society
William W. Sturm, History Room, Oakland Public Library
Barbara and Robert W. Foley, Madelyn Helling, D. Robert Paine,
      and Edwin L. Tyson, manuscript readers.

Prologue

For many years statuary played an important part in the commemoration of world figures and historical events, but by the end of World War I, the public lost interest in monumental sculpture in preference to the abstract movement that followed.

One statue that still commands attention is the Pioneer Memorial at Donner Lake in Northern California. Erected in honor of the immigrants who passed that way enroute to the new state, its site was chosen in memory of one of those emigrant trains, "The Donner Party."

Bogged down by snows the winter of 1846-1847 in the Truckee Pass, 84 of the original 90 men, women and children with ox-teams and canvas-covered wagons faced their last huge obstacle, the Sierra Nevada Mountain Range. At least half of them died and the others, mostly children, were carried out by relief parties made up of men who journeyed across the mountains to reach the starving immigrants.

Honoring them caused another human struggle within the committee responsible for the monument's erection; the play for power, egotistical demands and frustrating situations, balanced off by admiration, loyalty and dedication. This is the story of a notable achievement, encompassing a period of 20 years, by a group of individuals who made every effort to work together.

The Origin of a Statue

At the eastern end of Donner Lake, stands the impressive statue of a pioneer and his family. With hand-shaded eyes, the trail-weary emigrant gazes toward the mountain summit that must be crossed to reach the California valleys. His strong right arm supports his wife and baby, while an older child holds tightly to his boots.

This inspiring sculpture culminated a dream of Charles F. McGlashan, a Truckee attorney and author of *The History of the Donner Party*. He had interviewed many of the survivors of that tragic wagon train, and from the time of its publication, wanted a memorial to their memory. He spoke of such a tribute whenever the occasion arose, even attempting to raise funds himself. The project, however, needed the sponsorship of a large organization.

Frank M. Rutherford, Principal of the Truckee Public School, authored a Donner Monument resolution that was presented before the Grand Parlor of the Native Sons in Nevada City (1898), which resulted in the appointment of a committee to study the Donner memorial.

At the Grand Parlor of 1901, Dr. Chester Warren Chapman, a thirty-seven year old Nevada City dentist, was chosen chairman of the Donner Monument Committee, and an expenditure of $5,000 was authorized to be raised among its subordinate Parlors.

The dogmatic set of Chapman's jaw matched the squareness of his five-foot, six-inch frame. Sandy complexioned, with penetrating blue eyes, he was known to speak sharply when opposed and said to have had the tenacity of a bull-dog and the crustiness of a bantam-cock. But, for all these brash characteristics, he had a gentler side, too. He was a man of great courage, foresight and vision, with a deep respect for the bravery, strength and achievements of the pioneers. His untiring efforts, energy and unswerving determination, kept him in as

Chairman of the Donner Monument Committee until the completion of the work. Without this man, the inspiring Pioneer statue on which today's visitors gaze in awe and wonder would not be in existence, or C. F. McGlashan's dream realized.

Chapman set about to thoroughly acquaint himself with the subject, and wrote to McGlashan for his book. Profoundly affected by the story of the Donner Party as related on those pages, an overwhelming emotion surged through his being. He had lived and mingled among such pioneers all his life and knew the type of men and women they were. Nothing could deter them, no horror or fate appall them. On they came, thousands upon thousands, the sturdiest and finest the nation could afford. The subject grew upon Chapman until he became determined that this project would develop into what his organization, The Native Sons of the Golden West, had meant in perpetuating the memory of their forefathers.

To assist in financing the memorial, McGlashan offered his book to be republished and sold under the auspices of the Na-

Charles F. McGlashan, a lawyer and newspaper editor, in the latter part of the nineteenth century was the leading citizen in Truckee, California. His book, *The History of the Donner Party,* was published in 1879. Interested in all civic affairs, he initiated the idea of a Donner memorial.

2

tive Sons, he retaining the copyright. As souvenirs, he suggested a project instigated some years before, when fragments of the last decaying log of the Murphy cabin were collected and placed in small vials. There were 5,000 of them and his offer was readily accepted.

Chapman began raising the $5,000 among the subordinate Parlors by sending circulars for pledged amounts, but returns averaged $10.00. A few Parlors raised $100.00 and two or three managed $200.00. At the end of two years, a balance of $1,774.50 had been accumulated.

Along with the pledges came suggestions. One brother felt that the larger amounts should come from those nearest Donner Lake. "They'd be the only ones to receive any benefit from a monument there!"

Another wanted a monument of the stone found near Donner, one side to be polished, for a suitable memorial. "This should not cost such an enormous sum," he said, "In fact $2,000 ought to erect any desirable and creditable memorial."

Dentist Chester Warren Chapman was appointed chairman of the Donner Monument Committee despite never having been a member of the Grand Parlor of the Native Sons of the Golden West. His most active committee consisted of two Grand Vice-presidents, four past Grand Presidents, and two Grand Trustees.

Joseph R. Knowland, Editor of the *Oakland Tribune* and President of the California Historical Landmarks League, suggested making the monument plain but substantial, thus bringing down the cost to a reasonable figure. By so doing, he felt that the money would he easier to raise.

Chapman became irritated and curtly sputtered, "I have been trying for years to convince the men in our Order that whatever is done under its auspices should emulate the high standards of our organization. It seems to me quite incredible that you who have taken the work so seriously will be difficult to change into this way of thinking."

Men of influence and prominence did not deter him to do battle whenever opinions differed from his own. San Francisco's Mayor James D. Phelan drew equal fire. Phelan wrote to Chapman, "A granite shaft inscribed with a brief history would be a sufficient memorial when you consider that it will be located in an inaccessible place where very few will see it."

Chapman caustically replied, "The ideas you advance in your letter are some of the many that make it almost impossible to accomplish this work. There are some who even feel that a pile of boulders rolled from the hillside would be sufficient to mark the spot."

Then came the San Francisco earthquake and fire of 1906, and during the reconstruction period, the Donner Monument Committee remained in limbo, all except the doctor, who constantly thought, envisioned and planned for a suitable monument at Donner Lake. His ideas gradually took form and he talked them over personally with the most influential men in Grand Parlor and received their support.

When Joseph Knowland became the 1909-10 Native Sons Grand President, Chapman confronted him. "I propose that the monument shall stand for all the pioneers who crossed the plains, and consist, roughly speaking, of a lone figure standing with the right hand shading his eyes, and his gaze fixed upon the west. The height of the pedestal on which it shall stand to be 22½ feet, the height of the snow that winter of 1846-47." His

4

ideas for the height of the pedestal remained consistent through the years as did the pioneer figure with hand-shaded eyes.

Fred H. Jung, the Grand Secretary, strongly favored the erection of the monument, but felt the name Donner Monument Committee a misnomer if the monument was to be one to the pioneers. "It should certainly be something of magnificence," he said, "and it seems to me the only way in which the Order can succeed is procuring such a monument." The men of Chapman's generation were children of the pioneers, and the concept of a monument to honor their fathers and mothers had an overwhelming effect.

H. C. Lichtenberger, Grand Vice-President, wrote to the doctor, "I have read with deep interest your opinion for the monument. Nothing could be more beautiful, sentimental or inspiring than the memorial you propose to properly commemorate the heroic valor of our pioneer ancestors, but in my opinion the monument should be erected in San Francisco."

The Donner Lake area in 1910 was still quite primitive and remote. The demands of auto travel were unknown, and only a county dirt road wound its crooked way over the summit, around the northeast side of the lake and on to Truckee.

Knowland had added to Chapman's committee Frank Rutherford, who in 1910 was a Truckee attorney. He proved an alert and capable assistant. Knowland pushed for a completed monument by the time Grand Parlor was to meet that summer on the north shore of Lake Tahoe.

Rutherford, a member of Donner Parlor in Truckee, wanted to attract Grand Parlor delegates to the site if for nothing more than the laying of a cornerstone. "The most essential thing is to get it started," he told Chapman. "If Grand Parlor gets interested, it will do more toward raising the full amount required than anything else we can do."

It was just the incentive the doctor needed. Not only the cornerstone would be placed, but the heavy foundation to support the statue he had in mind. He had traversed the site the year before with his friend Charles McGlashan and knew just

5

where the monument should stand. McGlashan knew the exact location of the cabin sites at the lake as pointed out to him by some of the survivors.

Artifacts such as broken dishes and cooking utensils had been excavated, and old fireplaces located. The Murphy cabin had stood against the big granite rock in today's Tamarack grove; the log duplex belonging to the Graves and Reed families to the east of it, and the Breen cabin site on which the monument was to be erected, near the county road.

Joe Marzen, local butcher, who owned the land and used it for cattle grazing deeded the one acre which covered the Breen cabin site to McGlashan in 1893. Being disappointed in Dr. Chapman's turn of emphasis by making the memorial a Pioneer monument, McGlashan, on offering the one acre to the Native Sons, called attention to the deed, which expressly stated it was to be used for a monument to the Donner Party.

Saturday, May 11, 1910, was chosen as the day for the ground-breaking ceremony. Among the group were Charles McGlashan, Dr. Chapman and the Honorable Frank W. Rutherford, now a State Assemblyman. Four of the party were stationed at the corner stakes of the deeded acre, and by sighting across the center point, the proper spot of the Breen cabin was located. This was the original Schallenberger cabin built in 1844 and occupied by the Breen-Keeseburg families in 1846.

McGlashan was chosen as the one most appropriate to make the initial stroke. As he bared his head and stood with eyes uplifted to heaven the others instinctively uncovered theirs. In an impressive manner of speech, he dedicated the soil to the men, women and children who constituted that martyred party of pioneers.

C. A. DeLong, master craftsman for the O'Neil Granite Works, was chosen to lay a foundation 20′ x 21′ x 6′ for the monument. Donner Parlor's Committee hustled up tools and lumber needed for the construction. Sixty yards of Donner Lake gravel, 37 yards of granite rock and a supply of water had to be hauled to the site and mixed with 110 barrels of cement to form the base. Chapman had ordered the cement from Cowell Cement Company, whose bid had been a few cents lower than that of George A. Legg, Nevada City merchant, lodge brother and friend. No one was more surprised than the overly confident Legg himself. From eight to fourteen men worked daily, not counting the teamsters with teams. Suitable laborers proved a problem, and a few railroad hangers-on were enlisted to help. Hard liquor had to be furnished them as an inducement to work, and this whiskey bill forwarded to the Native Sons Finance Committee required an explanation.

The 3½-foot-square block of Raymond granite, to be laid as a cornerstone and manufactured at O'Neil's San Francisco factory, had a space reserved for it on the north side of the concrete base so that it faced the county road. On its polished side was the inscription, "This cornerstone marks the site of one of the Donner Party cabins where a monument will be erected under the auspices of the N.S.G.W. to the pioneers who crossed the plains."

Susan Alexson, daughter of Frances Donner Wilder, read in a local paper of the impending monument and wrote to Dr. Chapman. He replied, "The unveiling of the completed monument will not take place for several years, and as we cannot tell what may happen in even weeks or months, it would be better for the survivors to come this June if they can. We will lay a cornerstone and have suitable exercises on the site of the old 'Schallenberger' (Breen) cabin." Through her he learned of the names of the nine living members of the Donner Party, and sent invitations to the five Donner sisters: Elitha Donner Wilder, Bruceville, Sacramento County, and Leanna Donner App of Jamestown, Tuolumne County, who were daughters of George Donner by a previous marriage; and to Frances Donner Wilder, Byron, Contra Costa County, Georgia Donner Babcock, Walla Walla, Washington, and Eliza Donner Houghton of Los Angeles, daughters of Tamsen and George Donner. There was just one of the Breen family surviving, Isabelle Breen McMahon of San Francisco. Of the Pike family only Naomi Pike Schenck,

The Dalles, Oregon, and the two James Frazier Reed daughters, Virginia Reed Murphy, San Jose, and Martha (Patty) Reed Lewis of Santa Cruz were invited.

The erection of such a monument had not yet been sanctioned by Grand Parlor when it convened on June 6, and Chapman's greatest challenge lay before him in convincing the brothers of a $25,000 memorial and the means with which to finance it. This was to be done with help from the State Legislature and the Native Daughters through sales and individual donations. He was never at a loss for words and stressed the point nearest their hearts, that of being sons of pioneers. In this, he was deeply sincere. "We are worthy sons of worthy sires," he reminded them. "It is our mission to perpetuate the memory of our forefathers. That does not mean simply to mark spots where things occurred, but to tell the story. We cannot hope with a simple shaft to present the character of the California pioneer to men who shall live ages hence. In a few years it would arouse only curiosity."

The cornerstone with planned inscription and weighing 600 pounds arrived by freight on June 6, the opening day of Grand Parlor, and dedicated after its adjournment on Friday, June 10. Donner Parlor procured Rapp's Band of Truckee, arranged for the ceremonies and festivity, and had as their special guests, three of the Donner survivors. They came by train to visit the site where 64 years before as little children, they had endured never to be forgotten hardships. They were Frances Donner Wilder, Virginia Reed Murphy and Patty Reed Lewis. The eastern end of Donner Lake is a sacred spot where the tall pines sigh a requiem for the dead whenever a soft breeze stirs. The very presence of three survivors of that horrendous ordeal lent pathos to the solemn ceremonies.

The dedication ended with a barbecue in the tamarack grove amid the stumps of trees that were below the snowline when the weakened pioneers hacked them down, and where the Murphy cabin once stood. The vials of log fragments from that cabin were sold as souvenirs, bringing $250 to the fund.

To prove that a towering memorial was in the offing, the Donner Monument Committee dedicated its base on June 10, 1910. The open center was covered with boards and held chairs for the guests. Dr. Chapman stood beside the 600-lb. cornerstone and extolled the project to a sparse audience.

Dr. Chapman, in reflecting on the haste in constructing the foundation said, "I sweat blood during that week or ten days, for I had the main issue of establishing the monument on a broad basis in the Grand Parlor. I hate failure, and want to carry this through successfully, but I also want to come out of it alive, which accounts for the little scrapping I indulged in to get a living chance."

He had accomplished the completed foundation, the approval of Grand Parlor to proceed with building the monument, and the appointment of the committee he wanted by the incoming Grand President, Dan Ryan.

The vials of wood fragments were sold at $1 a piece during the July Fourth celebration in Nevada City and the three-day festivities held in San Francisco on September 9th to commemorate California's 60th year as a state.

Virginia Reed Murphy and Martha (Patty) Reed Lewis accepted the invitation to participate in the Admission Day Parade on Market Street as guests of the Donner Monument Committee. A large banner on the open barouch in which they rode conveyed their identity to the public. The survivors, although hampered by failing health, were most willing to assist as best they could in financing the Donner memorial.

Through James Phelan Chapman learned the name of San Francisco's leading sculptor, Douglas Tilden, who had been deaf and speechless since childhood. His work included a statue of Father Junipero Serra in Golden Gate Park and, most famous of all, the monument to mechanics at the intersection of Market, Bush and Battery streets.

Phelan also suggested the De Rome Foundry in San Francisco as the best for casting the monument. Tilden and De Rome were definitely interested in the creation of the statue. The idea so strongly appealed to Louis De Rome that he offered his work and materials at cost. It was through him that the doctor had learned of the size needed for the foundation to hold the statue he had in mind.

Tilden quoted a price of $8,000 for a single figure and $15,000 for a group. Chapman wanted a well-known artist and, being assured of support by the men of his committee, began a working basis with Tilden. He wrote to him, "Personally, I believe that you would be preferred to any other artist, and if you will make some concession, I can then hope to be freed from any influence that might urge for open bids."

He made note of the location as being 150 feet from the railroad, with passenger traffic of 367,000 a year, in the midst of one of the grandest resort areas on the Pacific Coast, the proportions of which in another half-century could only be imagined. All to bring lasting fame to the sculptor. With his letter, he sent a copy of McGlashan's *History of the Donner Party*.

Tilden completed reading the account of the tragedy and wrote to the doctor, giving his concept of an appropriate memorial: "I agree with you that the pedestal should be of rough granite. As to statuary, I am strongly inclined to give it a much more phonetic or story telling quality than a single figure can impart. I would like to depict actual privation so that posterity can see and understand. On top I would place the lone pioneer figure you described. At the bottom I would place all around in the shadow of the rest, a dozen or so figures sitting, lying, creeping in different attitudes from cowering anxiety to resignation of death."

Chapman could foresee trouble. His story was being sacrificed to art. Was there no master great enough to combine the two? Greatly disturbed, he wrote to Tilden, "While I can see all of the elements in the experience of the Donner Party, yet my concept of the Pioneers is that of a type of men who possessed courage, determination and endurance to suffer physical and mental strain, remaining steadfast and resolute to triumph over difficulties under which others succomb. I would want the figure that surmounts this pedestal to show by attitude and mien that he had experienced terrors and strain. I would want the eager, searching gaze to show a realization that the goal was

John McQuarrie's model for the statue of Father William McKinnon, Chaplain of the First California Volunteer Infantry, that stood in Golden Gate Park. The statue portrayed the image of strength, courage and determination, and it influenced the Donner Monument Committee's selection of McQuarrie.

near, and I would like the face to show the light of the conqueror's soul that never fails except through death, no thought of which could be gleaned from the steadfast gaze toward the promised goal. I would not have those who leave its presence exclaim that it was beautiful. I would have them square their shoulders and say, 'I CAN AND I WILL!'"

Tilden could see but one sombre theme, that of shadow, sadness, submission, resignation and death.

Chapman and his group finally agreed to put the monument design out to bid, and Tilden wrote, "If you should decide to proceed by competition, I regret that under no circumstances would I submit a design merely to show what I can do. A committee should not tell the sculptor to incorporate its ideas in the design as was the substance of much of our correspondence. Give the artist absolute freedom as to conception." And so, Douglas Tilden bowed out.

During the remainder of 1911 Dr. Chapman and his committee visited the studios of three sculptors, that of M. Earl Cummins, Arthur Putnam, and John McQuarrie. Each was presented with a copy of *The History of the Donner Party* and invited to submit ideas for the monument. The committee, however, was greatly influenced by a model that McQuarrie had made for a statue in Golden Gate Park of Father William McKinnon, who in 1898 was Chaplain of the First California Volunteer Infantry. It had the robust, determined attitude that the doctor wanted for his pioneer figure. It portrayed the courage needed to overcome defeat, and the confidence to conquer the mountain.

Cummins made a caustic observation of the McKinnon statue in a San Francisco newspaper. Chapman, surprised and chagrinned, commented, "It smacks of professional jealousy. I am curious to know if our leaning towards McQuarrie could have been the cause. It may be that Cummins can point to technical error from the professional artist's standpoint and also that the figure does not look like McKinnon, but there is inspiration in that figure, Cummins or no Cummins."

Forty-year-old John McQuarrie, who did the Bear Flag monument in Sonoma, was chosen among the three to sculpt the statue for the Donner Lake monument on May 19, 1913.

In writing to Charles McGlashan of the Committee's decision, Dr. Chapman added, "If your book had not been written, I firmly believe this monument would have been one of the many to pioneers which stand in some park near a populous center in various states telling no particular story, proclaiming no particular type of being and perpetuating no particular principles or traits of character for the leaders of men. You have clothed the history of an event, from the thought of which men shrink in horror, with a sentiment which arouses all their compassion and forces them to see the goodness and admire the nobleness of their souls."

With the sculptor agreed upon, Chapman again focused his attention to financing the monument, and encouragement for funds came from some unexpected sources.

A bill for the appropriation of $5,000 to the Donner Monument Fund passed both legislative branches of the state, and was one of the last bills to be signed by Governor Hiram Johnson during his first term in office.

In November 1913 the Truckee Chamber of Commerce, in preparing for a winter carnival, erected an Ice Palace at the cost of $10,000. According to W. G. Gelatt, Special Representative, the festival would have everything that made the winter carnivals at Montreal so famous. Charles McGlashan had agreed to their reprinting his Donner Party story in a large edition, the profit from which was to go to the Donner Monument Fund. Gelatt assured the Doctor that they only wished to widen the sale of the book and would also be willing to sell the vials of wood fragments without any monetary reward.

The Nevada County Board of Supervisors agreed to a subscription of $500 for the memorial fund, and the Native Daughters of the Golden West pledged $2,000. Fred W. Bradley, State Mining Engineer and former Nevada County resident, made a commitment of $1,500, payable as needed.

The doctor, elated, began searching about for other well-heeled and affluent individuals who might give a donation. His first quarry was James W. Phelan, United States Senator and former San Francisco mayor. He attempted to contact him personally, but that failing, wrote a five-page letter recalling their initial meeting at a Native Sons banquet, and then explained the goals of the Donner Monument Committee. In a more direct quotation, he wrote, "Knowing the interest that you have taken in matters of this kind, I have chosen to ask you to meet the expense of printing and binding one of the Donner Party editions in the sum of $1,500.

After a month when no word came from Phelan, Chapman sought the solace and advice of his colleague, John F. Davis.

"Possibly your comment at the banquet might have had something to do with the silence," Davis wrote. "However, if any mischief was done by it, there is no need of thinking about it now. I suggest that in a matter of this kind very little headway can be gained by correspondence and that the only thing to do would be to take up the matter of the subscription with Mr. Phelan personally the next time you are in the City. I imagine that the little incident of the banquet is long passed. He is too big a man to let anything of that kind rankle."

By the end of the second month, Phelan did answer. "I do not feel like complying with your request to finance an edition of the book at this time. I will, however, buy ten copies of the $5.00 edition when they come off the press." This touchy episode ended the doctor's direct and personal appeal for individual contributions.

Among the many irritations that plagued the Donner Monument Committee was the cancellation of Chapman's railroad pass. It took a special bill before the State Legislature to allow him free passage to do "patriotic work," a phrase it added to the list of non-paying travelers. By 1916, state laws were superseded by national authority and the Southern Pacific Company could no longer extend freight and passenger privileges. Before this final decision, an offer was made for free transportation to

Chapman's Committee when they gathered together to inspect the statue in San Francisco.

They met at McQuarrie's Studio (1370 Sutter Street) in March, and the sculptor's model was more than anticipated. True, the Committee knew of McQuarrie's addition of a family to the lone pioneer, but the four-foot group in molded clay held them in awe and reverence for the creative genius of this talented and unassuming man. The main figure depicted the strength of one who could blaze a trail west and hew the way; one who would throw himself before dangerous onslaughts to shield and protec those he loved. McQuarrie agreed to furnish one plaster pioneer group, sixteen feet high for the sum of $4,000.

"I had thought the artist would ask from $5,000 to $10,000 for his work," Chapman said, "but it is evident that McQuarrie figured on his labor and expenses only and asked nothing for his talents. I must confess to a pang of regret for this as I have become attached to him because of the faithful, intense interest he has taken. He has worked for two years, trying everything I wanted him to express and studying every trait of character that he was asked for without the guarantee of a cent. His hands and face were a perfect quiver as he watched us deliberating on whether he should do the work or not. I would like ultimately to put a bonus on the finished work if we feel that to future generations it will be an inspiration."

Chapman reported to Grand Parlor in March of 1915: "We will have a four-foot plaster model of the monument which the committee will offer to the Grand Parlor with the recommendation that it be placed somewhere in the N.S.G.W. Building as a permanent exhibit."

John Davis, whom Chapman relied upon as his San Francisco representative, called on McQuarrie in April and found him working on the plaster cast. "Of course you know that the clay statue is all destroyed in the process," he wrote. "I confess to a sentimental feeling of loss in finding the sturdy old fellow and his winsome spouse gone. However, that can't be helped.

This four-foot-high family group, modeled in clay by John McQuarrie, won the approval of the monument committee. Although Dr. Chapman had favored a lone pioneer, he readily agreed that the addition of a family gave it greater significance.

17

The figure stands out far more strikingly in the plaster than in the original clay. McQuarrie's man is cutting and whittling things down so that in a couple of days it will be finished as far as the plaster cast is concerned."

Then, on June 23, Davis informed the doctor, "I have just returned from a visit to the studio where McQuarrie has two men working on the large plaster statue. The work is going to be enormous. He is making fine headway but is financially in bad shape. It is impossible for him to borrow without making sacrifices that he should not be called upon to make. After what I have seen this morning, I would not hesitate in getting to him the necessary $500.

During the 1915 Panama Pacific Exposition in San Francisco, Chapman attempted to have the four-foot plaster model placed in front of the California Building on an eleven-foot pedestal of imitation granite, to be unveiled on September 9, California's Admission Day. However, the plan failed, causing him frustration and keen disappointment. If the doctor could have been on the spot, he would have seen it through, but directing its progress from far away Nevada City proved a futile effort.

Clarence Hunt, editor of the *Grizzly Bear*, who had been publishing news of the coming event in the N.S.G.W. official magazine, attempted to console Chapman. "You will find, doctor, as I have, that very little dependence can be placed in the ruling element of our San Francisco members. Their main purpose in the order is to extend their political power, and their assistance is given only when they think that purpose can be advanced. You have simply been imposed upon, just as I have been many times. We are both a long way from San Francisco, so it's easy to 'slip it to us' as the saying goes."

Louie De Rome gave a bid of $11,350 for casting and finishing the monument in bronze, a price that included assembling and placing the statue on its pedestal after it had been unloaded from the freight cars. He had conferred with McQuarrie as to the length of time required to make the plaster model and cast the bronze work, and both agreed it would take two years. Payments were to be made on a semiannual basis covering a period of four and a half years and a down payment of $2,500.

When De Rome visited McQuarrie's studio in August of 1915 he was most enthused with the progress. "It looks simply great," he wrote to the doctor. "I don't think that there has been a finer piece of statuary turned out anywhere. The lower half of the group is practically finished with the exception of the base which still has to be made. The upper half of the mother is nearly finished and on the man, the finishing layer of plaster is being applied."

Moving the model to the De Rome Foundry in Oakland became a problem. Due to the size of the statue teamsters refused to bid. Finally the Alpine Wood and Supply Company of Berkeley made an offer for $150. This included removing the back wall of the studio, loading the parts on a truck, replacing the wall, hauling the parts to the Oakland Foundry, and there to unload with the assistance of the foundry workers. By the end of 1916 the statue was in the shop and ready for bronzing.

The following year was made notable by the erection of the 22-foot pedestal on which the statue was to stand at Donner Lake. Dr. Chapman's representative at Truckee was Richard Falltrick, who collected the materials, took care of the hauling and contracted J. W. Woods to build the pedestal.

Rock and gravel came from around Donner Lake and lumber from the Truckee Lumber Yard. Charles McGlashan furnished the pipe to carry water from a spring and the Postal Telegraph Company provided temporary electrical power.

The concrete foundation erected on the Donner Lake site in 1910 became the base for the pedestal. Sufficient rock, some pieces the size of a man's head, had to be hauled to the monument site by team.

Woods wrote to Chapman on September 14, "We are a little over halfway with the work now. When you were here last, you said something about leaving a projection at the top of the pedestal, same to be made of the flat rock we had on the

This base for the pioneer statue was constructed of rock and gravel from around Donner Lake. Completed in September 1917, the height of the pedestal (22 feet) depicted the level of snow at Donner Lake in the winter of 1846-47. Treasured pioneer memorabilia is sealed within the aperture behind the bronze tablet.

ground. It is going to be a difficult job to do, and I would suggest that we make the projection of concrete and mark it off to imitate stone. Let it project two or four inches and rise five or six inches. That would make a concrete slab over the whole top." This projection brought the total height of the pedestal to 22½ feet, the height of the snow in 1846.

A small cubicle in back of the nameplate was reserved to receive the archives. Chapman assigned Charles Belshaw of Contra Costa County the task of securing a lead box, and to John Davis, the collecting of the archival materials.

A six-foot cube to serve as a water tank was placed in the top of the pedestal, and a small gasoline pump with a lift of thirty feet anchored to the original foundation, kept the tank full of water with which to irrigate the ground surrounding it. To reach this, a trap door was needed at the base of the pedestal, and fittings for a hose placed underground (not visible today).

Conduits and wire extended to the corners of the statue's base so that reflected light would bring it out in bold relief. For permanent power Chapman considered the Southern Pacific Company as a source. The liqhted statue would be seen from the cars and add to the company's siqhtseeing Donner Pass route. However, Falltrick measured the distance and found it to be 2½ miles. By putting in poles 100 feet apart, nearly 150 would be needed, so another source had to be acquired.

The pedestal stood completed by the end of September, 1917. The rocks had been cleansed of morter and gave the appearance of having been placed there dry.

John McQuarrie accompanied Chapman to Truckee where he designed a finish to the top slab so that the feet of the statue would come into view from the proper point and the cornices proportioned in the balance.

As to the bronze work, De Rome reported, "The heads and arms are cast, and the two lower limbs of the man will be ready for the oven in a day or two."

When completed, the statue assembled at the plant was searched for imperfections and became a source of curiosity to excursions of local school children. Then it was carefully crated in sections and made ready to be transported by train to Truckee.

The six-by-eight-foot bronze tablet, to be placed in front, needed an inscription conveying the purpose of the sculpture. To Dr. Chapman there was no one who could do justice to this other than Benjamin Ide Wheeler, President of the University of California. He would be able to give those few words the exact shade of sentiment, and John Davis was again commissioned to contact Wheeler.

Davis met with the University's president, gave him the dimensions of the tablet and explained the type of wording that would reflect the attitude of the pioneers, their courage, steadfastness, etc. "He was very gracious about it all," Davis wrote, "but I will have to let him have a description and photograph of the monument."

In the spring of 1918 Davis wrote to the doctor for a finer shading of the nameplate inscription, and Chapman answered, "My idea is to perpetuate the 'Spirit' of the pioneer, that self-reliant character which takes the initiative and presses on to success, prompted only by desire and courage from within himself, looking nowhere for reward but in the achievement of the purpose. That character which voluntarily takes the brunt of life and lets who may reap benefits, looks to no earthly source for support or protection, looks no where for praise or for censure and heeds none. I believe our statue will stimulate and create strong human impulses, that it will arouse determination to meet life squarely, face on and honestly. I appreciate that much is asked of President Wheeler. We want added inspiration, perhaps in the form of an apostrophe or paraphrase of the attributes which place one in the van of life. The monument itself is the tribute."

President Wheeler, inspired, wrote to Chapman, "Your letter to Judge Davis gave me just the impulse I needed. As you will see from the enclosed, I have made ample use of your wonderful characterization of the spirit of Pioneers. By gathering all your fine and enthusiastic flashes of light together, I saw the illuminated picture of the whole as you intended it. Many thanks. I have given great care to the composition and the subject was worthy of any care any man could offer. At best my words may be inadequate. I hope they will not disappoint you too much."

Enclosed in the letter were the following poetical lines:

> Verile to risk and find
> Kindly withal and a ready help
> Facing the brunt of Fate
> Indomitable—Unafraid

De Rome promised to have the bronze nameplate ready, and a smaller one the doctor ordered to read, "In commemoration of the Pioneers who crossed the plains to settle California. Erected under the auspices of the Native Sons of the Golden West and the Native Daughters of the Golden West. Dedicated June 6. 1918."

Joe V. Snyder of Hydraulic Parlor #56 in Nevada City was elected Grand President for the term of 1917-18. A resolution was made to hold the next Grand Parlor at Truckee in the Grand President's home county, where the dedication of the Donner Monument would be a part of the program.

Truckee's Chamber of Commerce and the people of the town began making plans to host the delegates. Dances and theatricals were held to raise funds for the occasion. Besides the dedication of the $35,000 monument, a trip to Lake Tahoe was in the offing.

However, another obstacle loomed before them when James Lick Parlor of San Francisco adopted a resolution calling upon the Parlors of the Order to petition the Grand President for a change of meeting place from Truckee to San Francisco. The reason being that the country, now at war, was calling on everyone to economize.

John Davis reacted. "This country was at war at the time of last Grand Parlor, and to say now that we did not know of what we were doing, is such a pitiful confession of incompetency, that I do not understand the argument."

Chapman wrote and answered letters. "I had a hard time to keep up with my correspondence without the move to change the Grand Parlor meeting place." His efforts were successful, as the subordinate Parlors rallied around the Donner Monument Committee and the Truckee location. One more hurdle towards the completion of the monument had been conquered.

Although all the survivors of the Donner Party were invited to the dedication, all but three declined due to ailing health. Naomi Pike Schenck of The Dalles, Oregon, sent $500 to Chapman's Committee, but being one of the pioneers to be honored, it was not accepted. The doctor suggested that he be allowed to present her gift to the Red Cross. After the dedication he wrote, "I have 136 feet of film showing the principle features of the program which I will send to you if a picture-house there will run it through." (This film was made by Gaumont and Company, San Francisco.)

On May 1 Chapman contacted John Davis. "Another matter has just come up which might provide additional interest in the meeting of the Grand Parlor and inspire some feeling of patriotism. My son, Allen, has just returned from Call Field, Texas, with a Lieutenant's Commission in the Aviation Corp. He is on a ten-day furlough, six of them being necessary for traveling. He enlisted at nineteen and is now twenty. Brother Snyder immediately suggested that Allen attempt a flight over the Sierra during the Grand Parlor session. Allen is eager, but he is in the service and that would require Army orders. It would also require an army machine and perhaps a couple of mechanics. If such orders were issued in the interest of recruiting, and for arousing patriotism, spreading it about the State, it strikes me that it would pay for the service itself. A successful flight over the Sierra by an Army aviator under Government orders where attempts by others had always failed, would be a fitting setting for the ceremony of dedicating such a monument."

Although Dr. Chapman did not instigate the flight, he be-

At twenty years of age, the serious face of young Lieutenant Allen Chapman peers out from under a bulky World War I flyer's helmet. Eager to make the first successful flight over Donner Summit on Dedication Day, he studied air currents and emergency landings.

came one of its ardent and enthusiastic supporters. He encouraged Senator Ingram to gain the support of Governor Stephens, and to interview Senators Johnson and Phelan.

Grand President Joe V. Snyder telegraphed President Woodrow Wilson, "Honorable John P. Davis is special envoy to interest you in an endeavor to stimulate patriotic interest in Government appeals. Your recognition would greatly enthuse our people. Flight accomplished over Sierra by Government would reassure public mind considerably in west. We are heartily with you."

Snyder received a reply from Congressman John E. Raker: "Chapman will be detailed to make flight if possible to spare him. If not, some other Californian will be assigned."

The doctor was elated. It meant another historical achievement. After the failure of the Donner Party to scale the mountain heights, the trains with their silver palace cars wound their way through tunnels and snowsheds along the mountain sides, but now as the monument was being dedicated, the first airplane to cross the summit would appear. What an accomplishment, and Dr. Chapman wanted the honor for his son. He hurried off a telegram to the officer-in-charge at Call Field:

"Conditions here make Lieutenant Allen Chapman of greater patriotic influence in flight over Sierra than any substitute. Would appreciate endeavor to spare him."

"Allen assured me," Chapman wrote to the editor of the *Grizzly Bear*, "that he can make the flight successfully if the Government gives him a ship that will make the lift and he says that they have ships that will do 10,000 feet. He wants time to select emergency landing places, and to try to cut air currents with the airplane in which he is to make the try. I would have preferred to have been more modest so far as my part is concerned, but Joe scoffed at it, and Governor Stephens insisted that I should not be held responsible for instigating it. He is supporting it heartily."

However, the officer-in-charge at Call Field felt differently, and Davis wrote, "My dear Doctor: I am extremely disappointed for your sake at the final outcome in the effort to have Allen Chapman detailed and an airplane assigned. The Military Department turned it down. So encouraged was Congressman Raker at the first interview he wired that the matter was arranged, but overnight the whole situation changed and Colonel Bane, himself a Californian, familiar with the ground, turned it down cold. Judge Raker interviewed him in person, and tried to get him to reconsider, but it was no use."

The resolute character that was his forbid Dr. Chapman to admit defeat or show disappointment. He bit his lower lip, held his chin a little higher, but his face appeared greyed and the shoulders drooped.

The statue began arriving in crates at the Truckee station on May 19, and Henry Lichtenberger of Donner Parlor wrote, "We secured tackle, blocks, ropes and cables from Pendergast, the Southern Pacific railroad foreman here. He has given us his expert advice and without asking him, consented to inspect the pole and appliances before they start to make the lift. We are surely fortunate to have a man with his experience in handling wrecking cranes to offer his assistance. Everything has worked out so far without a hitch of any kind. They expect to start in tomorrow to haul. My vocabulary is too limited to express the grandeur of your work, and I am carried back to the time in Oroville when you said, 'Henry, I could put up an affair costing $10,000, but I am not going to do that. When that monument is completed, it will be a credit to our State.' It surely is!"

Five days later he wrote, "The monument is nearing completion. The child and father are in place, and this morning they began to hoist the mother to her resting place. It should be completed by tomorrow."

A pillar reinforced with a steel rail was inserted into the larger figure for support, and on May 25 the 18-ton statue stood gazing across the wide expanse of mountain range. De Rome cemented the 1200-pound tablet into place, covering the niche containing the lead box of memorabilia the day before its dedication.

THE PIONEER MONUMENT
AT DONNER LAKE

Three survivors of the Donner Party, Martha "Patty" Reed Lewis, Eliza P. Donner Houghton, and Frances E. Donner Wilder stand in front of the monument on dedication day, bracketed by Nevada Governor Emmett D. Boyle (left), and Governor William D. Stephens of California (right).

A platform, 12 x 50 feet, large enough to hold 100 people was constructed 50 yards from the statue and near the road, permitting the audience to stand in front of the monument, but far enough away to get the best view. Lighting was arranged and the trees trimmed so that it could be seen from the railroad below the summit.

Six hundred Native Sons began arriving in Truckee on June 5, 1918, fully expecting the time to drag, but found this 41st Grand Parlor one long to remember. As guests of honor at Tahoe Tavern, they were given a warm welcome by Charles F. McGlashan. The steamers *Tahoe, Nevada,* and *Lake Tahoe* were pressed into service to convoy 500 of them around the Lake. Those who were fishing enthusiasts could be found on the Truckee River, Donner Creek or nearby Independence Lake, and the trout banquet held in town proved to be an evening of good fellowship.

The main event, however, was the dedication of the Pioneer Monument on June 6. Between 3000 and 3500 people attended the ceremony, many coming from the state of Nevada. Three hundred automobiles filled the parking area. Dr. Chapman presided, and speeches were alternated with music by Native Sons bands from Nevada City and Grass Valley.

Principal speaker Governor William D. Stephens of California shook hands with Nevada's Governor Emmett D. Boyle, symbolizing friendship between the two states. Others among the speakers were Dr. Henry Morse Stephens of the University of California History Department (representing President Wheeler), and the Grand Presidents of the Native Sons and Daughters, Joe V. Snyder and Grace S. Stoermer. (This marked the first time in Grand Parlor history that a Native Daughter addressed Native Son delegates.)

Guests of honor were three of the Donner Party survivors, Martha (Patty) Reed Lewis, Eliza Donner Houghton and Frances Donner Wilder. John McQuarrie and Louis De Rome were introduced.

Charles McGlashan gave the dedicatory address and the

great moment came when two little girls, Helen Chapman and Kathryn (Betsy) De Rome, drew the veil from the monument.

Surprise and delight enveloped the onlookers as they stood with eyes uplifted to the golden bronze shining in the sun. The figure of the man gazing towards the west, and that of the woman with babe in arms and garments flowing gracefully about her, depicted strength, courage, protectiveness, love and faith. Those attending its dedication were viewing a true artistic achievement, and the applause, impulsive, loud and clear, resounded across the Truckee meadows.

In 1919 the Grand President contacted Chapman regarding defects in the monument that needed immediate repairs, and suggested a fence be erected around it. The fact that a caretaker was needed became imminent, and when T. C. Wohlbruck, a San Francisco photographer, wrote for permission to set up a souvenir and refreshment stand near the monument, his request couldn't have been made at a more opportune time. Wohlbruck offered to care for the monument, erect a fence and sell the remaining vials of wood fragments. He already had two "Service Canteens," as they were known—the first at Emigrant Gap and the second on the summit above Myers Station (Highway 50) near Lake Tahoe.

Grand Parlor owned the one acre on which the monument and canteen stood, but Dr. Chapman, envisioning Donner Memorial Park, felt the size inadequate. To Louis De Rome, he commented, "If things go well and it seems opportune I will try to get a bill through the Legislature to park the site of the monument. I am planning now for a large park, at least 150 acres, to include campgrounds. They are right in curbing me if I can be curbed!"

Through Judge Davis an agreement was made with the Pacific Fruit Express Company, which now owned the surrounding property, to buy eight acres at $50 each and an additional amount later on.

During the Grand Parlor session of 1918 held in Truckee, the Native Sons Historical Landmarks Committee was instructed

T. C. Wohlbruck, a San Francisco photographer, became the first monument custodian and in 1920 he built the lodge at the right for a souvenir and refreshment stand. His plan to build rental cabins, a rustic archway entrance and a stone fireplace ended in 1928 when the monument and grounds were presented to the State of California.

by resolution to mark the three Donner Party cabin sites at the lake, and Chairman Joseph E. Knowland wrote to Chapman the following year: "Sometime next week a tablet will be placed on the Murphy rock. The Breen tablet is complete, but in placing the Graves tablet, a piece of granite for the base is needed. I wondered if there would be any objection to our using the original cornerstone of granite which stands near the present monument."

It wasn't the unused cornerstone that infuriated Dr. Chapman, but the Breen tablet to be placed on the east side of the monument. The inscription began "On this spot stood the Breen Cabin of the party who started for California from Springfield . . . ."

To the doctor, the naming of any pioneer lost the intent of the monument, which was to symbolize the characteristics of those pioneers who settled California. His determination to avoid this special recognition to any one person was reflected in a statement to "Patty" Reed Lewis. "As you know, the monument near Donner Lake is a tribute to all pioneers who crossed the plains to settle California. It was placed where it is because the 'Donner Party' was in our opinion the most typical of all those early emigrant trains. We refrained from a special mention of that party in tablet-form lest it be construed as a memorial of the tragedy which befell them rather than the type of beings they exemplified. We also avoided the mention of any individuals lest it place the others in the background."

To Charles McGlashan, however, the monument retained its original purpose, that of honoring the Donner Party, and the differences of opinion between the doctor and himself caused the deterioration of their friendship, and became one of the most volatile controversaries Grand Parlor ever had.

Most disturbing was Chapman's insistence that the monument was not built on the Breen cabin site. He had made no mention of it until after Knowland's Committee announced its intention of marking the three cabin sites. He seemed genuinely in earnest, and to Grand Parlor he said, "I wish to make a

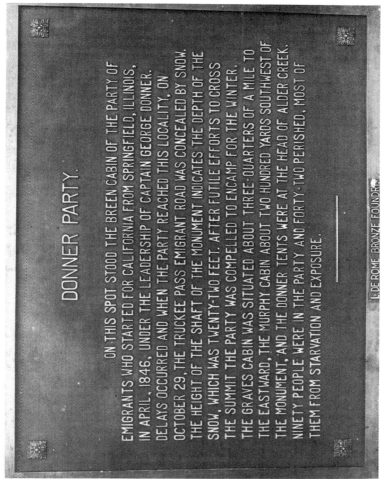

The original
version of the
Donner Party
plaque.

sworn statement that the monument stands approximately fifty feet from the spot which General McGlashan designated as the site of the Breen cabin just before breaking ground for the foundation in 1910."

In McGlashan's testimony (a ten-page printed document) he emphasized his knowledge of the location. "If any man on earth knows the site of the Breen cabin, I do, and I am ready to take an oath that it is covered by the foundation of the monument. . . . Year after year, I have pointed out the site to visitors. . . . Since 1879, it was marked by the stump, the excavated floor, and by the fireplace, for I did not dig up the fireplace in searching for relics. The ashes were sifted, but the burned earth and foundation of the fireplace stood undisturbed . . . . Anyone could find the site of the Breen cabin from 1879 to 1910."

Dr. Chapman contacted C. A. DeLong, who had laid the foundation for his opinion of the ground's surface. "It would be hard to tell whether any excavation had been made," replied DeLong. "Personally I think not as we found granite rock large and small, mixed in with loose sand and dirt such as is common in the vicinity."

Chapman contended that McGlashan once stated that he kept an accurate account of the location of the Breen cabin site until 1910, and had since carried with him only the impression that the committee erected the monument at the spot which had been pointed out by him. This McGlashan denied.

They met at Donner Lake one May morning in 1920 and began digging. Some flat stones were found and a couple of hand-pressed bricks that could have been used by campers for a campfire. Nothing was proven, but Dr. Chapman won the decision when the Committee on Historic Landmarks recommended the adoption of a resolution containing Chapman's sworn statement. The tablet on the east side of the monument now begins with "Near this spot . . ."

The decision greatly affected the attitude of Charles McGlashan. To have the tablet read "Near this spot stood the Breen cabin," when he knew the monument was on the exact spot, was to him an injustice to history, He never visited the monument again.

Louis De Rome had to wait out a payment due to this confrontation. The Finance Board contended that enough money had been appropriated for construction of the monument but had been used for a purpose that had no bearing on the construction itself, namely to settle a controversy as to the exact position of the Breen cabin!

The cabin site remains unmarked, and as for the Graves cabin, a white cross placed there by Charles McGlashan was later moved by an ice company to a point further north and on higher ground to make way for an ice pond.

In 1923 Dr. Chapman was relieved from his work as committee chairman. William J. Hayes, Grand President, wrote, "I am writing to advise you that your name was left off the Donner Monument Committee only because we decided that Grand Parlor Committee appointments should be restricted to members of the Grand Parlor. I talked the matter over with members of the Board of Grand Officers and that was the conclusion arrived at, and the rule followed throughout. I sincerely regret that I will have to forego the pleasure of naming you to serve on the Donner Monument Committee during my administration, for I know that you are deeply interested in the work of the order and its future welfare."

The Pioneer Monument, presented as a gift to the State of California in 1928, became an Historical Point of Interest in 1931, and a National Historic Site on September 9, 1962. It was then that 2000 or more people gathered to witness the dedication of the $250,000 Emigrant Trail Museum, including the Harold T. (Bizz) Johnson Audio-Visual Room, named for the congressman who played an important part for fifteen years to bring it into existence.

The Pioneer Monument has since been dwarfed by the construction of Highway 80. Built in 1967, it cloverleafs nearby with approaches to Reno, Truckee and Lake Tahoe. The statue, clearly seen, attracts many curious visitors, but Dr. Chapman's struggle to keep it as a memorial to all pioneers has lost significance throughout the years. Standing today on that tragic site, it seems destined to be known as the Donner Monument.

Dr. Chester W. Chapman stands in front of the historical masterpiece that has become a tribute to our heritage.

Appendix A

THE MONUMENT SPECIFICATIONS

Location: Two miles west of the town of Truckee, Nevada County, California, and near Donner Lake.

Sculptor: John McQuarrie of San Francisco.

Cost: $4,000.

Nameplate: Weight, 1,200 pounds. Inscription composed by Dr. Benjamin Ide Wheeler, President of the University of California.

Casting: By the Louis De Rome Brass, Bell and Bronze Foundry, Oakland, California. Cost, including nameplates, $12,000. After the death of Louis De Rome Sr., Louis De Rome Jr., continued the work and provided the materials at cost, as promised by his late father.

Measurements: *Base:* 21 x 20 x 6 feet. *Pedestal:* Approximately 15 feet square and 22½ high, the height of the snow that winter of 1846-47. *Statue:* Male figure, 16 feet high. *Weight:* 18 tons.

Cost: $35,000.

Archives: *Placed behind nameplate:* Dedication memorabilia; mementos from survivors of the Donner Party; copy of *The History of the Donner Party* by Charles F. McGlashan; vials of wood fragments from the Murphy cabin; Native Sons papers and newspaper clippings; copies of *The Grizzly Bear Magazine*; newspapers relating to the event; photos of progress in construction; list of those who donated funds or services; coins.

Dedicated: June 6, 1918, by the Native Sons of the Golden West and Native Daughters of the Golden West.

Appendix B

Committee members, appointed each year by the incoming Grand President, changed from year to year (1900-1928), and many Native Sons served. They were so scattered in Northern California it became necessary for Dr. Chapman to assume full leadership, and he resorted to calling on them whenever needed.

Those who remained active throughout the years were: Frank N. Rutherford, Henry C. Lichtenberger, and Richard Falltrick of Truckee; John P. Davis and Lewis P. Byington, San Francisco; Charles M. Belshaw, Contra Costa County; and Clarence E. Jarvis, Sutter Creek.

Appendix C

FACSIMILE SIGNATURES OF THE
DONNER MONUMENT COMMITTEE OF 1918

Bibliography

BREEN, PATRICK. *Diary of Patrick Breen*, University of California, 1910.

CHAPMAN, C. W., Collection of Donner Monument Committee letters, 1901-1923, Searls Historical Library, Nevada City, California.

HOUGHTON, ELIZA P. DONNER. *The Expedition of the Donner Party and Its Tragic Fate.* Los Angeles, 1920.

MCGLASHAN, CHARLES F., "The History of the Donner Party." San Franciscc, California, 1931.

MCGLASHAN, M. NONA. *Give Me a Mountain Meadow.* Valley Publishing, 1977.

MURPHY, VIRGINIA REED. "Across the Plains in the Donner Party, 1856." *Century Magazine*, July 1891.

PADEN, IRENE D. *The Wake of the Prairie Schooner.* McMillan, 1945.

REED, VIRGINIA. Letter dated May 16, 1847, written at age 12 to inform eastern family of the Donner tragedy. Bancroft Library, University of California.

TESTIMONIES:

Dr. C. W. Chapman and C. F. McGlashan to the 44th Annual Session, California Grand Parlor, Native Sons of the of the Golden West, 1921.

PAMPHLET:

"The Donner Party Tragedy." California State Division of Beaches and Parks.